Growing Up

Seed to Pumpkin

by Sonia W. Black

EXPLORE the LIFE CYCLE!

Content Consultant

Rosie Lerner, Program Assistant
Purdue Extension Horticulturist
Purdue University

SCHOLASTIC

Library of Congress Cataloging-in-Publication Data
Names: Black, Sonia, author.
Title: Seed to pumpkin/Sonia W. Black.
Description: New York: Children's Press, an imprint of Scholastic Inc.,
    2021. | Series: Growing up | Includes index. | Audience: Ages 6-7. |
    Audience: Grades K-1. | Summary: "Book introduces the reader to the life
    cycle of a pumpkin"— Provided by publisher.
Identifiers: LCCN 2020031787 | ISBN 9780531136935 (library binding) | ISBN 9780531136997 (paperback)
Subjects: LCSH: Pumpkin—Life cycles—Juvenile literature.
Classification: LCC SB347 .B53 2021 | DDC 635/.62—dc23
LC record available at https://lccn.loc.gov/2020031787

Produced by Spooky Cheetah Press. Book Design by Kimberly Shake.
Original series design by Maria Bergós, Book&Look.

Printed in Heshan, China 62

SCHOLASTIC, CHILDREN'S PRESS, GROWING UP™, and associated logos are
trademarks and/or registered trademarks of Scholastic Inc.

1 2 3 4 5 6 7 8 9 10 R 30 29 28 27 26 25 24 23 22 21

Scholastic Inc., 557 Broadway, New York, NY 10012.

Photos ©: 1 grass and throughout: Freepik; 9, 10-11: redmal/Getty Images; 12-13 vine: Chansom Pantip/Getty
Images; 17 bottom: Subas chandra Mahato/Getty Images; 18: Inga Spence/Science Source; 21 fungus leaf: Nigel
Cattlin/FLPA/Minden Pictures; 24: Andersen Ross Photography Inc/Getty Images; 26 center right: Photo courtesy
of the Deerfield Fair Association; 27 top left: Fiantas/Getty Images; 27 center right: panida wijitpanya/Getty
Images; 29 flowers: Scott Camazine/Science Source; 29 young pumpkin: John Kaprielian/Science Source.

All other photos © Shutterstock.

Table of Contents

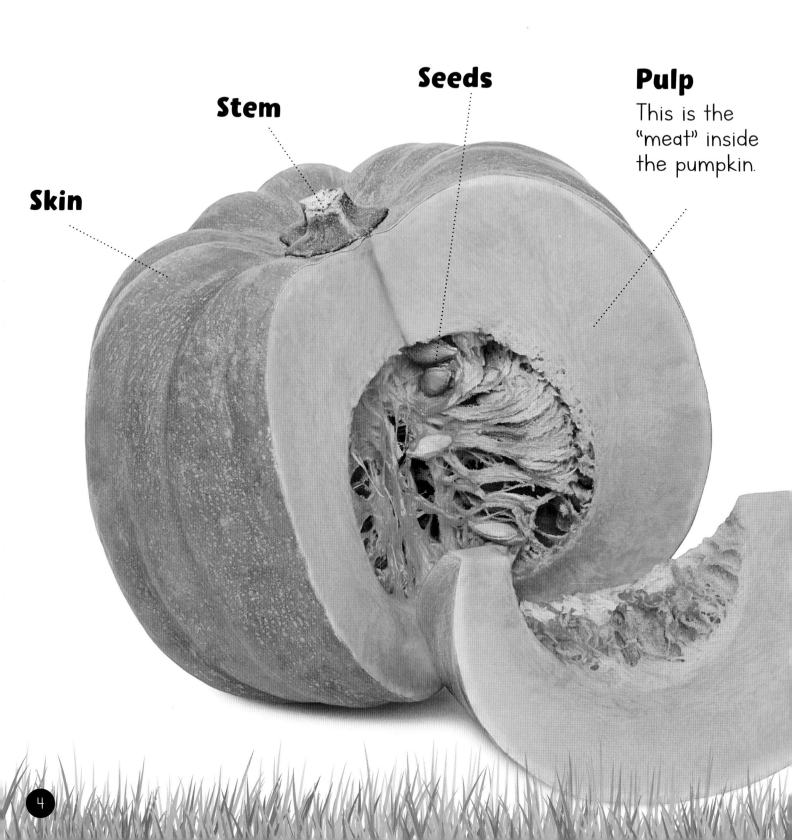

Stem

Seeds

Pulp
This is the "meat" inside the pumpkin.

Skin

What Is a Pumpkin?

Many people think a pumpkin is a vegetable. But it is a fruit! Like all fruits, a pumpkin grows from a flower. It has seeds inside. There are more than 45 different kinds of pumpkins. And they all start out the same way: as a tiny seed.

The pumpkin seed is planted about one inch deep in the soil. ▶

It Starts with a Seed

A pumpkin seed is planted in the pumpkin patch. The seed needs warm soil to grow. It will not grow in cold weather. It will not grow when there is **frost** on the ground. The soil has **nutrients**. The soil has rainwater, too. The nutrients and water help the seed grow.

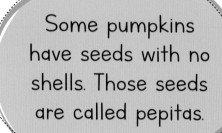
Some pumpkins have seeds with no shells. Those seeds are called pepitas.

Taking Root

The hard outer shell of the seed is called the seed coat. The seed takes in nutrients and water through the seed coat. In a few days, the seed cracks open. A big root and a stem poke out. The root grows down, down into the soil. The stem grows up and up. It pushes the seed coat toward the sky.

Soil

Seed Coat

Roots

Stem

A Sprout Grows

Soon little roots grow from the big root. They have important work to do. They collect more nutrients and water. They feed themselves and the stem, too. The stem keeps growing up through the soil. In about a week, it bursts through the top. It grows into a **sprout**!

Some people plant pumpkin seeds just to eat the sprouts as veggies.

Sprout

A Vine Forms

The sprout needs energy to grow. Its little leaves use sunlight, air, and water to make food. This process is called **photosynthesis**. Each day, the sprout

grows into a bigger plant. Larger leaves grow along its stem. The plant grows outward along the ground. It is a leafy **vine**.

Other vine fruits include watermelons and grapes.

Male
flower

Female
flower

Many people like to eat pumpkin flowers.

Flowers Bloom

The leafy vine grows and spreads out over the ground. Soon, big, beautiful yellow flowers grow on the vine. Some are female; others are male.

Hungry bees fly onto the bright flowers. They are looking for **nectar** to eat. The busy bees also do a big job for the pumpkin plant.

Making the Fruit Grow

The male flower makes pollen, which gets stuck to the bees. The bees carry the pollen to the female flower. This is called **pollination**. Pollination makes the pumpkin fruit start to grow.

Soon the stem right below the pollinated female flower swells up. This is the pumpkin fruit growing. It looks like a little ball.

More than one-third of the world's **crops** rely on bees for pollination.

Pollen

Pumpkin fruit

Little Green Pumpkin

The female flower turns brown. The **petals** dry up and fall off. Now the little green ball turns into a little green pumpkin. It grows and grows on the vine.

Pumpkins need the right amount of water to grow. If it is very rainy, the pumpkin plant may get too much water. If dry weather lasts a long time, the plant won't get enough water. The pumpkin could die if the gardener doesn't water the plant.

Threats to the Pumpkin

Even if the pumpkin plant grows perfectly, it still faces threats. Some animals and insects like to feed on pumpkin leaves. The leaves may be damaged or become **diseased**.

Bugs like cucumber beetles and squash bugs chew holes in flowers and leaves.

Deer love to eat the fruit from the pumpkin plant.

Then they can't make food for the plant. Large animals might visit the pumpkin patch to eat the **ripe** fruit. Smaller animals like mice, squirrels, and chipmunks bite through the fruit to eat the seeds.

Diseases
can kill the plant.

Rabbits
nibble on the plant's stem and leaves.

The name pumpkin comes from the Greek word *pepon.* It means "large melon."

A Ripe Pumpkin

The plant grows into a pumpkin in about 80 to 120 days. By fall, the little green fruit isn't little anymore. And it's probably not green, either. Now it is a ripe orange pumpkin. The green vine is all dried up and brown. The pumpkin is ready to be picked.

Pumpkins of all sizes and shapes can be found in the pumpkin patch.

After the Pumpkin Patch

A pumpkin patch has loads of ripe pumpkins! Big pumpkins. Small pumpkins. Take your pick! Some people like to carve a jack-o'-lantern for Halloween. Others bake yummy pumpkin pies and cookies. They may even clean and dry some seeds and either eat them or store them away. When spring comes, they can plant their pumpkin seeds. The life cycle of the pumpkin will begin all over again!

Pumpkin Facts

Pumpkins come in different colors. Most are orange. But some pumpkins are yellow, white, green, or even red.

In 2018, a man in New Hampshire grew the largest pumpkin in U.S. history. The giant pumpkin weighed 2,528 pounds. That's heavier than a full-grown bull.

Jack-Be-Little (orange) and Baby Boo (white) are two of the smallest types of pumpkins. Some people use them for fall decorations.

In Ireland, people didn't carve jack-o'-lanterns out of pumpkins. They carved faces in turnips and beets (pictured) instead!

If you clean your pumpkin seeds and dry them completely, they can last for six years. Or you can bake them for a crunchy snack!

More than 1.5 billion pounds of pumpkins are grown in the United States every year. Illinois grows the most.

Pilgrims made pumpkin pies differently than we do. After removing the pulp and seeds, the cook filled the shell with milk, honey, and spices, then baked it.

Growing Up from Seed to Pumpkin

A pumpkin starts out as a tiny seed. It goes through many changes as it grows into a beautiful pumpkin.

Ripe pumpkin
The pumpkin changes color to orange when it's ripe and ready to pick.

Seeds
Inside the seed are a root and a sprout waiting to grow.

Sprout
At first, the pumpkin plant is tiny and has two leaves.

Vine
The sprout grows into a vine with lots of leaves.

Flowers
Flowers form on the vine.

Young pumpkin
After pollination, a green pumpkin starts to grow.

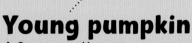

Glossary

crops (KRAHPS) plants grown for food for people or animals

diseased (di-ZEEZD) affected with a sickness

frost (FRAWST) a fine layer of powdery ice that forms on surfaces when the temperature goes below freezing

nectar (NEK-tur) a sweet liquid from flowers that bees gather and make into honey

nutrients (NOO-tree-uhnts) substances that promote growth and maintain life

petals (PET-uhlz) the colored outer parts of a flower

photosynthesis (foh-toh-SIN-thuh-sis) a chemical process by which plants use energy from the sun to turn water and carbon dioxide into food

pollination (pah-luh-NAY-shuhn) the process by which seeds are created through the transfer of pollen between flowering plants

ripe (RIPE) fully developed or mature; ready to be harvested, picked, or eaten

sprout (SPROWT) a new or young plant growth

vine (VINE) a plant with trailing or climbing stems that grows along the ground

Index

About the Author

Sonia W. Black is a former editor at Scholastic. She has worked on books in a range of formats, from picture books to young adult novels. Black is also the author of a number of books for early readers. She lives in a small town in New Jersey, where she enjoys pumpkin-picking time on local farms when fall comes around.